ideal
school supply

MW01714937

FUNtastic FROGS™

Numbers & Beginning Place Value

By Judy Goodnow
& Shirley Hoogeboom

Contents

Published by Ideal School Supply
an imprint of
Frank Schaffer Publications®

ideal school supply™

Editor: Jill Osofsky
Cover Concept and Page Design: Sara Mordecai
Cover Illustration: Kristin Mallory
Text Illustrations: Jim Connolly

Frank Schaffer Publications®

Ideal School Supply is an imprint of Frank Schaffer Publications.

Send all inquiries to:
Frank Schaffer Publications
8720 Orion Place
Columbus, Ohio 43240-2111

Funtastic Frogs™: Numbers & Place Values, Grades K–2
ISBN 1-56451-364-5

3 4 5 6 7 8 9 10 GLO 12 11 10 09

Teacher's Notes

This book is one in a series of books designed to develop children's mathematical thinking. Each book supports the use of Funtastic Frog™ Counters in activities that teach key mathematical concepts. This book introduces the Logs for Frogs™ to explore place-value concepts.

Funtastic Frog™ Counters and Logs for Frogs™ are available from Ideal School Supply or your local Ideal School Supply dealer. Please visit our Web site at *www.MHteachers.com.*

In *Funtastic Frogs™: Numbers & Beginning Place Value, Grades K-2,* children use the frogs and logs to:

- count and write numbers
- represent numbers in different ways
- explore place-value concepts for ones, tens, and hundreds
- represent numbers on a place-value chart
- use reasoning to solve problems

Exploring numbers with frogs and logs is an appealing introduction to place value in the base-ten number system. The frogs and logs provide both concrete and visual models to help children develop an initial understanding of place value and the base-ten number system. Each log is made to hold ten frogs. The logs can be snapped together to make a "raft" of one hundred. All the activities support current mathematics standards.

Contents

This book includes 22 activities divided into sections and one game.

Sample solutions are provided on page 7. A photocopy master for the frog counters and logs are included on page 8.

Math Skills and Understandings	**Related Activities**
Count and write numbers	Activities 1-6
Explore place-value concepts	Activities 7-10; Game
Explore tens and ones	Activities 11-18; Game
Explore hundreds	Activity 19
Represent numbers on a place-value chart	Activities 7-19
Use place value to solve problems	Activities 20-22

Suggestions for Classroom Use

By using the frogs and logs children develop the understanding that a numeral can stand for 1, or for 10, or for 100, dependingon the location of the digit in a number.

111

100 10 1

Putting 10 frogs on a log helps children begin to understand that 10 frogs can be thought of as 10 individual frogs (10 ones) or as 1 ten. Children need to be able to see that 12 individual frogs (12 ones) can also be thought of as 1 ten and 2 ones.

The activities are sequenced by level of difficulty. Modify a section if you find it is too challenging for your children or not challenging enough.

The activities can be introduced to the whole class using an overhead projector or sitting together in a small group. Once children understand the directions for the activities, they can work in pairs, small groups, or individually at a learning center.

Materials

You may wish to make an overhead transparency of the first activity page in each section as an introduction to the activities that follow.

For each child, pair of children, or group, you will need:
- pencils
- a tub of 108 small frogs
- 12 logs
- an activity sheet for each child or pair of children

Note that two dice numbered 1-6 are required to play *Tic-Tac-Frogs.*

Introducing the Activities

Before children begin doing the activities, give them time to just explore putting frogs on logs. This will give them time to satisfy their curiosity about the frogs and logs. You might also suggest that they snap some logs together.

While the children are exploring, you might ask questions such as, *Can you show me 8 frogs? Can you show me 12 frogs? How many frogs do you need to fill a log? If you take 15 frogs and fill a log with them, how many frogs are left over?*

When you introduce the activities in a section, lead the children through the first activity. Discuss the directions and how to record their work. The *Do More* part of each page encourages the children to apply what they have learned in the activity to solve a problem.

When reviewing the activity as a group, encourage the children to talk about their thinking. Children may use different methods to find an answer. Hearing different strategies to find an answer will provide different models for other children to consider. Class discussions encourage the type of thinking required for to develop good number sense, reasoning, and problem solving. It also provides the opportunity to use the language of mathematics.

Mathematical Content by Section

Activities 1-6: Count and write numbers. The children take frogs to match the number of frogs on the page, count the frogs, and write the number that tells how many. Later they add a number of frogs to the group then count and write the number that tells how many in all.

Activity 7-10: Explore place value and represent numbers in a place-value chart. The children fill one or more logs with frogs, then count the frogs. Later they fill logs, count the frogs left over, then find how many frogs in all. They record their work in a place-value chart.

Activities 11-18: Explore tens and ones and record on a place-value chart. The children take logs and frogs as shown, fill the logs with frogs, count the frogs left over, and then find how many frogs in all. In activity 13, the children begin calling a full log a *ten* and an individual frog a *one*. In activities 15-16 the children record the number a first way—the number of *ones*. Then they fill a log or logs with the frogs, and record the number a second way— the number of *tens* and *ones*. They record their work in a place-value chart for each activity.

Tens	Ones	Number of frogs in all

Activity 19: Explore hundreds and record numbers on a place-value chart. The children snap together 10 full logs to make a raft of 100 frogs. They call a ten-log raft a *hundred.* They show 100 in three different ways, then record in a place-value chart.

Activities 20-22: Solve problems. The children use clues about full logs *(tens)* and frogs *(ones)* to find mystery numbers.

Tic-Tac-Frogs: Children apply what they have learned about *tens* and *ones* and place-value concepts to play a game.

Sample Solutions

These are sample solutions. Additional solutions are possible for some activities.

Activity

1 A 5 **B** 8

DO MORE: Answers will vary.

2 A 9 **B** 12

DO MORE: Answers will vary.

3 15

DO MORE: The child will have more frogs.

4 20

DO MORE: Molly has more frogs.

5 13

DO MORE: Sara has more frogs.

6 17

DO MORE: Rachel has fewer frogs.

7 A 10

B 20

DO MORE: 30

8

	Full logs	Frogs left over	Number of frogs in all
A.	1	2	12
B.	1	9	19

9

	Full logs	Frogs left over	Number of frogs in all
A.	2	0	20
B.	3	1	31

DO MORE: 35

10

	Full logs	Frogs left over	Number of frogs in all
A.	2	5	25
B.	3	3	33

DO MORE: 48

11

Full logs	Frogs left over	Number of frogs in all
3	0	30

DO MORE: 59

12

	Full logs	Frogs left over	Number of frogs in all
A.	5	0	50
B.	5	8	58

DO MORE: Answers will vary.

13

	Tens	Ones	Number of frogs in all
A.	6	0	60
B.	6	3	63

DO MORE: 75

Activity

14

	Full logs	Frogs left over	Number of frogs in all
A.	7	0	70
B.	7	9	79

DO MORE: 5 tens and 0 ones

15

Tens	Ones	Number of frogs in all
1	6	16

DO MORE: 18 ones

16

Tens	Ones	Number of frogs in all
2	4	24

DO MORE: Answers will vary.

17

	Tens	Ones	Number of frogs in all
Way 1	6	5	65
Way 2	5	15	65

DO MORE:

	Tens	Ones	Number of frogs in all
Way 1	3	24	54
Way 2	4	14	54

18

	Tens	Ones	Number of frogs in all
Way 1	9	0	90
Way 2	8	10	90

DO MORE:

	Tens	Ones	Number of frogs in all
Way 1	2	35	55
Way 2	3	25	55

19

	Hundreds	Tens	Ones	Number of frogs in all
Way 1	1	0	0	100
Way 2	0	10	0	100
Way 3	0	0	100	100

DO MORE: 103 or 104

20 A 38 **B** 50 **C** 82

21 A 50 **B** 61 **C** 75

22 A 66 **B** 105 **C** 58, 60, 68, or 70

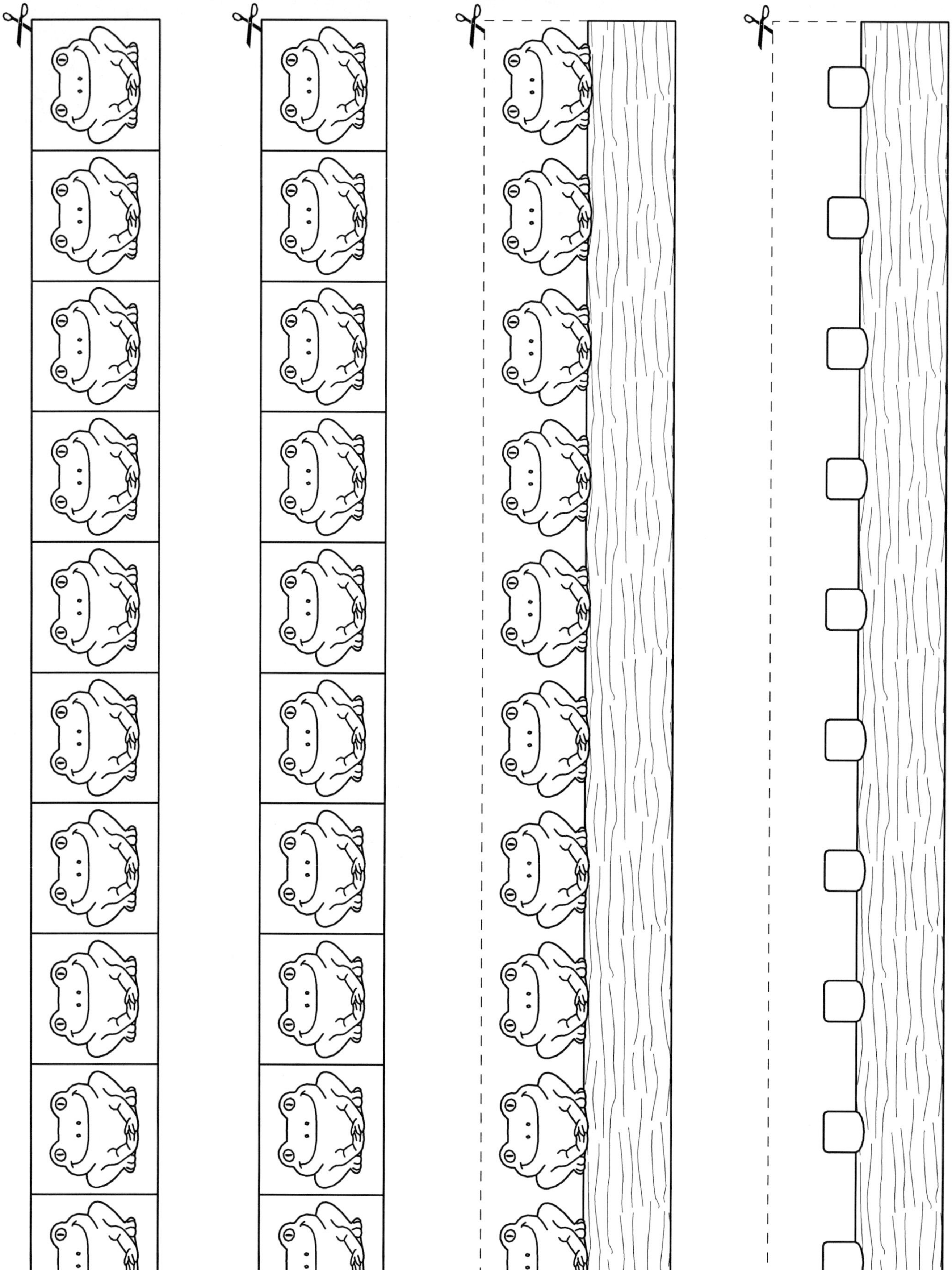

ACTIVITY 1

Name ______________________

Use small frogs.

A. Match each frog.
Count.

How many frogs? __________

B. Match each frog.
Count.

How many frogs? __________

DO MORE

How many frogs can you hold in one hand?

Guess. Then count. __________

Name ______________________

ACTIVITY 2

Use small frogs.

A. Match each frog.
Count.

How many frogs? __________

B. Match each frog.
Count.

How many frogs? __________

DO MORE

How many frogs can you hold in two hands?

Guess. Then count. __________

Count and write numbers

Name ____________________

Use small frogs.

Match each frog.
Count.

How many frogs?__________

DO MORE

William has 10 frogs.
How many frogs did you match on this page?

Who has more frogs?__________

Name ____________________

ACTIVITY 4

Use small frogs.

Match each frog.
Count.

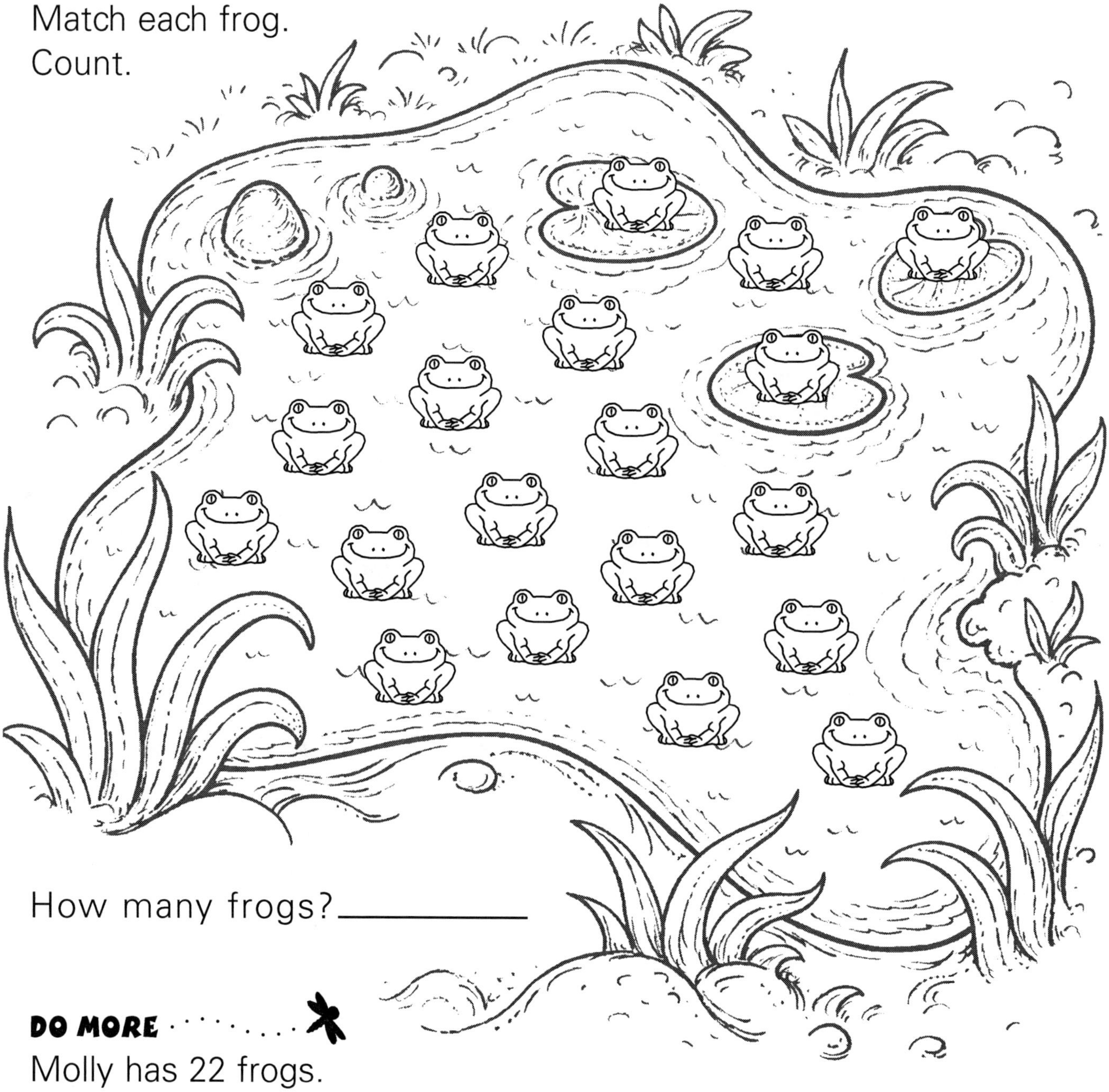

How many frogs? __________

DO MORE

Molly has 22 frogs.
How many frogs did you match on this page?

Who has more frogs? __________

Name ______________________

ACTIVITY 5

Use small frogs.

Match each frog.
Then add 4 more frogs.
Count.

How many frogs in all? __________

DO MORE

Sara has 20 frogs.
Joel has 19 frogs.

Who has more frogs? __________

Name ____________________

ACTIVITY 6

Use small frogs.

Match each frog.
Then add 6 more frogs.
Count.

How many frogs in all? __________

DO MORE

George has 33 frogs.
Rachel has 28 frogs.

Who has fewer frogs? ____________________

Explore place value

ACTIVITY 7

Name ____________________

Use small frogs and logs.

A. Match each frog.
Take 1 log.
Fill the log with the frogs.

How many frogs on 1 log? __________

B. Take 2 logs.
Fill the logs with frogs.

How many frogs on 2 logs? __________

DO MORE

To fill 3 logs, how many frogs do you need? __________

Explore place value

ACTIVITY 8

Name ______________________

Use small frogs and logs.

A. Match each frog.
Fill logs with the frogs.
Count and record.

B. Take 9 more frogs.
Fill logs with the frogs.
Count and record.

	Full logs	Frogs left over	Number of frogs in all
A.	1	2	
B.			

Explore place value

Name ____________________

Use small frogs and logs.

A. Match each frog.
Fill logs with the frogs.
Count and record.

B. Take 11 more frogs.
Fill logs with the frogs.
Count and record.

Full logs	Frogs left over	Number of frogs in all
A.		
B.		

DO MORE

Mary has 3 full logs.
She has 5 frogs left over.
How many frogs in all does Mary have? __________

Explore place value

Name ____________________

Use small frogs and logs.

A. Match each frog.
Fill logs with frogs.
Count and record.

B. Take 8 more frogs.
Fill logs with your frogs.
Count and record.

Full logs	Frogs left over	Number of frogs in all
A.		
B.		

DO MORE

Tommy has 4 full logs and 8 frogs left over.
How many frogs in all does Tommy have? __________

Explore tens and ones

Name ____________________

Use small frogs and logs.

Take logs to match.
Fill each log with frogs.
Count and record.

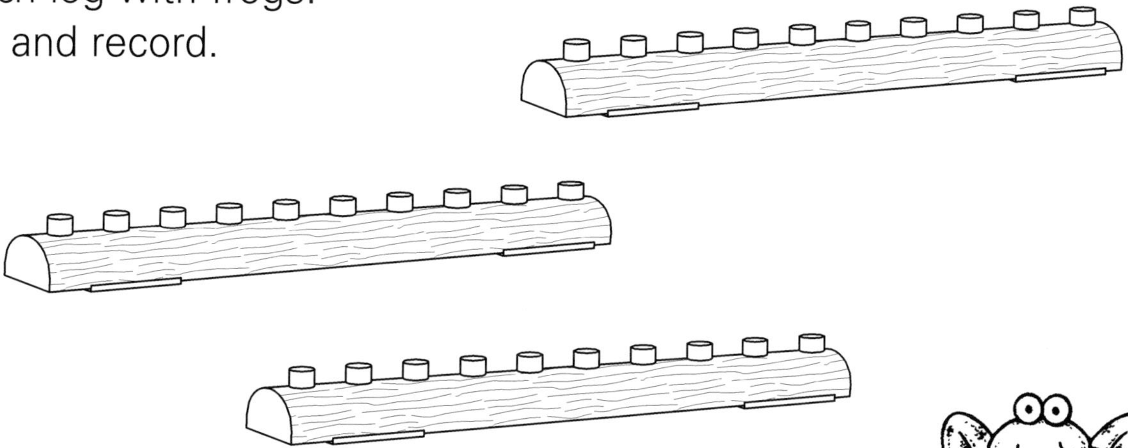

Full logs	Frogs left over	Number of frogs in all

DO MORE

Karen has 5 full logs.
She has 9 frogs left over.
How many frogs in all does Karen have? __________

ACTIVITY 12

Name ______________________

Use small frogs and logs.

A. Take logs to match.
Fill each log with frogs.
Count and record.

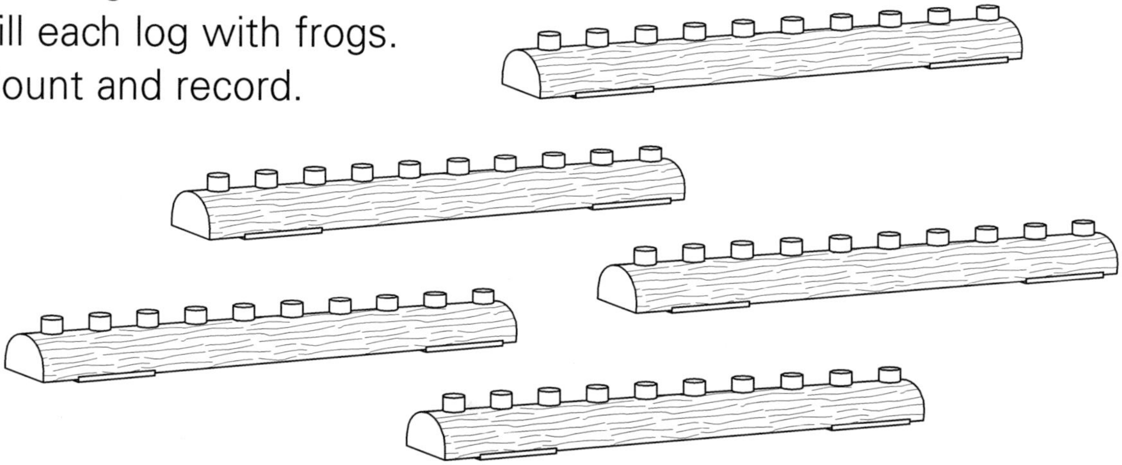

B. Take 8 more frogs.
Count and record.

Full logs	Frogs left over	Number of frogs in all
A.		
B.		

DO MORE

Write your own problem. Let someone else solve it.

I have ____________ full logs.

I have ____________ frogs left over.

How many frogs in all do I have? ____________

Explore tens and ones

Name ____________________

Use small frogs and logs.

A. Take logs to match.
Fill each log with frogs.
Count and record.

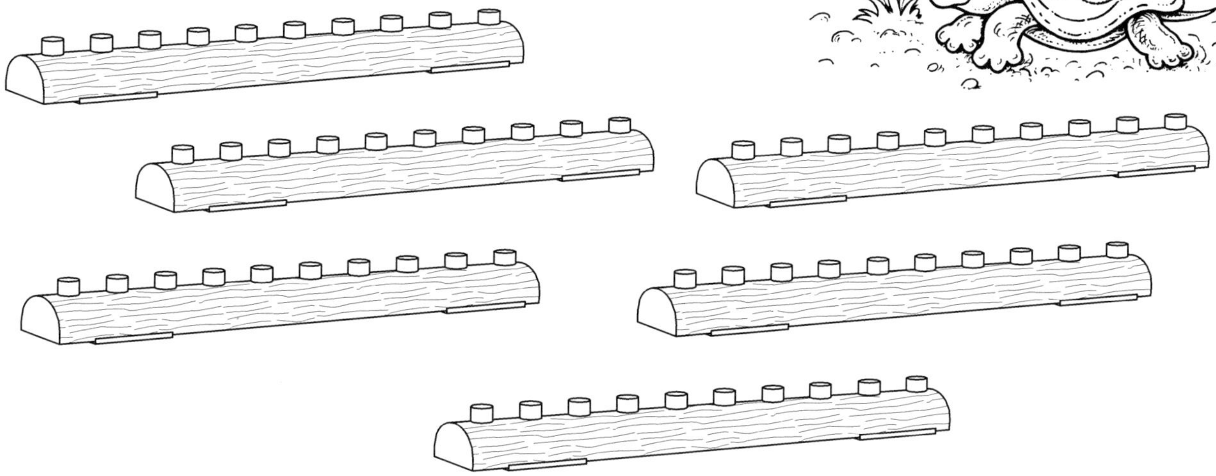

B. Take 3 more frogs.
Count and record.

Tens	Ones	Number of frogs in all
A.		
B.		

DO MORE

Sam has 7 tens.
He has 5 ones.
How many frogs in all does Sam have? __________

Name ____________________

Use small frogs and logs.

A. Take logs to match.
Fill each log with frogs.
Count and record.

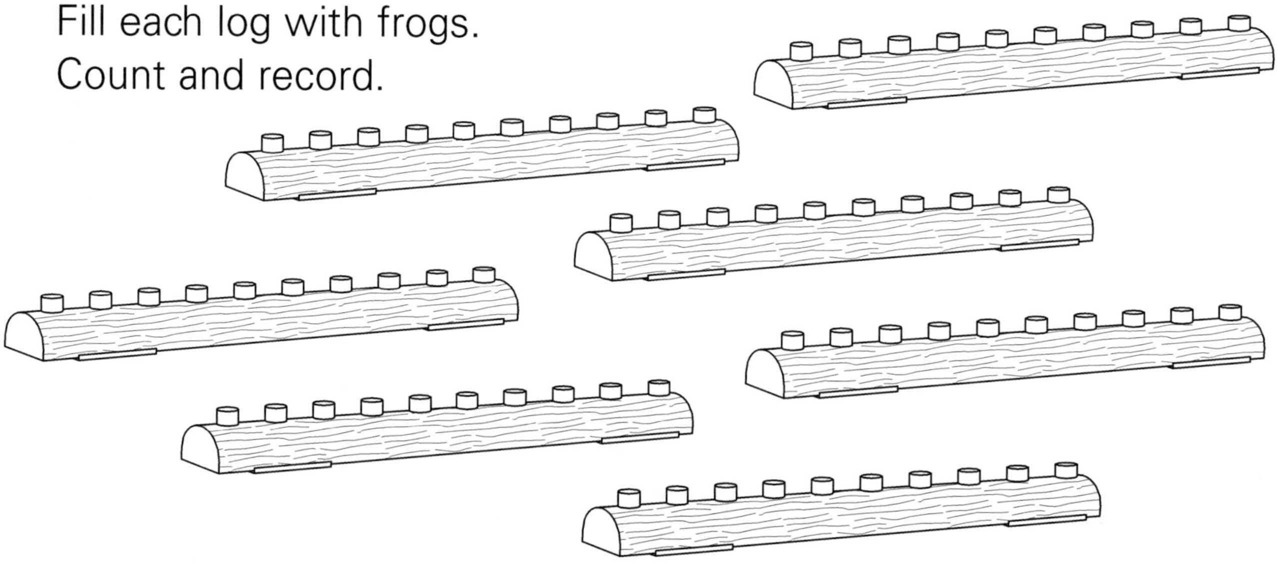

B. Take 9 more frogs.
Count and record.

Full logs	Frogs left over	Number of frogs in all
A.		
B.		

DO MORE

Jennifer has 50 frogs in all.
How many tens does Jennifer have? __________

How many ones does she have? __________

Explore tens and ones

ACTIVITY 15

Name ____________________

Use small frogs and logs.

A. Match each frog.
Count how many. __________

B. Fill a log with the frogs.
Count and record.

Tens	Ones	Number of frogs in all

DO MORE

Alex has 38 frogs in all.
He has 2 tens.

How many ones does Alex have? __________

Explore tens and ones

ACTIVITY 16

Name ______________________

Use small frogs and logs.

A. Match each frog.
Count how many.__________

B. Fill logs with the frogs.
Count and record.

Tens	Ones	Number of frogs in all

DO MORE

Write your own problem. Let someone else solve it.

I have__________frogs in all.

I have__________ tens.

How many ones do I have?__________

Explore tens and ones

Name ____________________

Use small frogs and logs.

Use frogs and logs to show 65 two ways.
Fill in the chart.

	Tens	Ones	Number of frogs in all
Way 1			
Way 2			

DO MORE

Fill in the chart.

	Tens	Ones	Number of frogs in all
Way 1	3		
Way 2		14	

Explore tens and ones

ACTIVITY 18

Name ______________________

Use small frogs and logs.

Use frogs and logs to show 90 two ways.
Fill in the chart.

	Tens	Ones	Number of frogs in all
Way 1			
Way 2			

DO MORE

Fill in the chart.

	Tens	Ones	Number of frogs in all
Way 1		35	55
Way 2		25	

Name ____________________

Use small frogs and logs.

Fill 10 logs with frogs.
Snap your logs together to make a raft.

What number does your ten-log raft show? __________
Show the number in three ways.
Fill in the chart.

	Hundreds	Tens	Ones	Number of frogs in all
Way 1				
Way 2				
Way 3				

DO MORE

Tina has 1 hundred.
She has no tens.
She has more than 2 ones, but fewer than 5.

What could Tina's number be? __________

Solve problems

ACTIVITY 20

Name ____________________

Use small frogs and logs.

Read the clues.
Find the mystery numbers.

A. Clues

I have more than 2 full logs.
I have fewer than 4 full logs.
I have 8 frogs left over.

What is my mystery number? ________

B. Clues

I have more than 3 full logs.
I have fewer than 7 full logs.
I have an odd number of full logs.

What is my mystery number? ________

C. Clues

I have more than 6 full logs.
I have fewer than 7 full logs.
I have 12 frogs left over.

What is my mystery number? ________

DO MORE

Write your own clues.
Let someone else find your mystery number.

Solve problems

Name ____________________

Use small frogs and logs.

Read the clues
Find the mystery numbers.

A. Clues
I can show it with 4 tens and 10 ones.
I can show it with 5 tens.

What is my mystery number?__________

B. Clues
I can show it with 6 tens and 1 one.
I can show it with 4 tens and 21 ones.

What is my mystery number?__________

C. Clues
I can show it with 6 tens and 15 ones.
I can show it with 5 tens and 25 ones.

What is my mystery number?__________

DO MORE
Write your own clues.
Let someone else find your mystery number.

Solve problems

Name ______________________

Use small frogs and logs.

Read the clues.
Find the mystery numbers.

A. Clues
I have more than 5 full logs.
I have fewer than 8 full logs.
I have an even number of full logs.
I have 6 frogs left over.

What is my mystery number? __________

B. Clues
I have a full ten-log raft.
I have 3 more than 2 frogs left over.

What is my mystery number? __________

C. Clues
I have more than 4 and fewer than 7 full logs.
I have between 6 and 12 frogs left over.
I have an even number of frogs left over.

What could my mystery number be? __________

DO MORE
Write your own clues.
Let someone else find your mystery number.

Name ____________________

Tic-Tac-Frogs

for 2 players

READY

You need 12 frogs—6 each of two colors.
You need the *Tic-Tac-Frogs* game board.
You need 2 dice numbered 1-6.

SET

Sit side by side.
Each of you take 6 frogs of one color.
Take turns rolling the dice. Add the numbers rolled.
The player who rolls the higher number begins.

PLAY

To win, you must have 3 frogs in a row in any direction on the game board.

Take turns. On your turn, roll the dice. Choose one die to show tens, and one die to show ones. Find a square on the game board to match your number. Put a frog on the square. Keep playing in the same way.

The first player to have 3 frogs in a row is the winner.

Name________________________

Tic-Tac-Frogs

Less than **20**	More than **10**	Less than **15**	Less than **30**
Less than **30**	More than **35**	More than **20**	Less than **40**
More than **40**	More than **5**	More than **25**	More than **50**